Urban Survival Handbook

The Beginners Guide to Securing your Territory, Food, and Weapons

(How to Survive Your First Disaster)

⚠ CAUTION

COPYRIGHT

Copyright © 2015

DISCLAIMER

TABLE OF CONTENTS

INTRODUCTION

When disaster strikes, the truth is that only those who are prepared for it survive. Mother Nature is a force that cannot be curbed by any man-made invention or technology, and she has the power to decimate everything in her path. In that moment, the calmer and more decisive you can be, the better the odds of your survival.

For most people, panic is the first emotion they feel when face to face with a disaster. Whether it is a hurricane, earthquake, or even a tornado, the essential rules of survival always stay the same. However, the vast majority of people are now living in urbanized centers and are heavily dependent on modern technology to sustain them. Basic survival techniques must be adapted and enchanted in order to better suit the environment most of us live in today. Another important point to consider is the human element. After a natural disaster — and this is especially true in urban areas — human reaction will be just as unpredictable as the disaster itself. In their panic, the populace can resort to looting and violence, possibly putting you and your loved ones at risk. In modern day society, not only will you have to protect yourself from the forces of nature, but also from the forces of your fellow man.

In this installment of the Urban Survival Handbook, the basics of securing your shelter, food, and weapons will be laid out for you. All three of these are quintessential elements of survival, and you must act quickly in the wake of an event to secure all of them. We will take these elements and apply them to the modern day setting that they will be enacted in. Having a shelter already made for you, in the form of a house or apartment, is an asset that is invaluable. This asset must be fully utilized for your survival, including the objects within it. Using items that can be found around your house, you will learn how to make sure that your shelter is properly fortified and defensible, how to collect, ration, and stockpile your food and water sources, as well as how to utilize everyday items as weapons in case of a hostile situation. After this first primer, you will understand the key components and groundwork of what you need to do in order for you and your loved ones to survive any disaster.

This will give you the confidence that, once a disaster strikes, you will be ready to act in a calm and decisive manner. It will also ensure you are taking the right steps to sustain your survival. Eventually, you will have to make a decision on what to do next. However, the steps in this guide will give you the ability to do so in safety.

SECURING YOUR HOME

After the initial disaster, you will have to work quickly to fortify your home against both nature and potential intruders. In a natural disaster, the initial event is usually not the only one that takes place. Earthquakes can have aftershocks that are just as devastating as the first quake; tsunamis can continue flooding areas even after the initial wave; hurricanes can continue violently for days on end, causing flooding and wind damage in their wake. It is extremely important that you start to work on fortifying your shelter as soon as you can. The more times your home is hit by these destructive forces, the more likely it becomes that you will have less and less of the original structure of your shelter. The steps you need to take in order to protect your home can vary from disaster to disaster. Obviously, a flood will have some different protocols than an earthquake, but the overall intent stays the same — making sure your home will be somewhere where you can either ride out the disaster, or plan your next move. Taking a survey of the house is extremely important after the disaster. You have to identify any weak spots or damage within the structure, and work rapidly to repair them.

The first thing you will want to do is turn off your gas, electricity, and water lines. During the disaster, it is extremely probably that some, if not all, of these utilities will be damaged. This can be extremely dangerous for you, as all three of these utilities can kill you in different ways. Therefore, after a disaster, the first step you will take is making sure all of your utilities are shut off.

Shutting Off Your Utilities

If there is a gas leak within the central piping of your house, you are exposed to an incredible risk that could lead to death. The slowest, yet surest, way to die within your home is carbon monoxide poisoning. The gas that we use in our homes for cooking and cleaning has a large amount of carbon monoxide in it, and once our bodies gain an elevated concentration of the chemical, they slowly begin to shut down. Symptoms of poisoning can include dizziness, headache, nausea, and drowsiness. If the level of carbon monoxide reaches critical levels, you will pass out and your organs will shut down until you inevitably die. Detecting carbon monoxide levels without a meter is an extremely difficult task. While gas in the air can certainly have a smell, it is not always strong enough to be detected by the human nose. Even the smallest leak can accumulate and slowly fill the house with this fatal chemical. That is why it is so important to immediately shut down the gas lines into your house. To do so, you will need to find the main gas shutoff valve. It is usually located in the front, back, or side of your house. If you live in an apartment, the main gas line to your house will usually be from the stove, which you can easily turn off by looking behind it. It is a good idea to learn the location of this valve as soon as you can, so you know exactly where to go to get to it. The utility box will usually have a meter somewhere on it, making it easily identifiable. Electricity must also be turned off in the event of a disaster, as it is possible that broken or flooded lines can spark, causing a fire which could spread quickly. To do this, and once again I recommend looking beforehand, you must locate your main circuit breaker. It will usually be a recessed cabinet usually with a metal door that is found somewhere within or outside of your home. Many homes have the circuit breaker located in the closet or garage, while others have it on an outside wall. Once you find the breaker, simply flip all of the switches to the "Off" position. This effectively breaks the circuits, and cuts off the electricity to your household. Lastly, you will shut off the main water line to your house. The reason for doing this is that in the event of a disaster or attack, the main water supply can become cross-contaminated. By shutting off this supply, you can ensure that whatever water you do have remaining in your house is still salvageable and safe to drink. It is difficult to ascertain whether or not the water supply has been tainted, as many chemicals are colorless and odorless, so it is best to cut off the supply immediately.

While you are going around shutting off your utilities, make sure that you are actively surveying your space. Awareness is an extremely important aspect of survival, and one that must be practiced with due diligence. You are looking for potential spots of weakness within the structure of your home. Actively note any holes, windows, and doors that you have, so that when it comes time to fortify them later, you will have somewhat of a mental outline.

Securing The Outdoors

Additionally, this is a good time to make sure the outside of your house is sound. Preferably, you will not be spending much time outside, exposed to the elements, so take advantage of this time. Look around for any materials that may be useful (we will go over what items are important later in the guide), if your car is parked outside and still in running condition, move it to the garage where it will be safe. In a survival situation, there is no such thing as being overcautious, as you cannot rectify what may occur if you are not. Always assume that the worst is going to happen, and prepare beforehand accordingly. In the event of you needing to escape your surroundings quickly, a car can prove invaluable, acting as both transportation and shelter. While outside, also make sure that your front yard is relatively clear of debris. You do not want to give possible assailants places to hide, or take cover should they try to invade your house. Ensure that you will have a clear line of sight all the way to the road in front of your house readily visible from a front window, and the same with the back.

Special Considerations

Before you continue, take a moment to remember any special considerations that you may need to make for anyone in your household during the days to follow. An extremely important concern to take care of right away is to secure any medication that you may have in your house. Often times, a person's life can depend on the medication that they take each and every day. To stop doing this during a survival situation can pose a myriad of threats, many of which cannot be dealt with directly. Grab all of your medication, and store it in a central location. Even over the counter medications, such as aspirin or Benadryl can prove invaluable should you need it. If you have pets, make sure that they are securely restricted to one room in the house. A disaster is not only stressful for you, but also for your pet. It will be frightened, and perhaps even skittish. The last thing you need during this first few integral moments is to be chasing after your pet, or for your pet to possibly escape whilst you secure the house. For the time being, put them in a room with the door closed.

First aid is another concern to take care of right off the bat. Chances are, that someone in your house will need first aid, if not now, then in the future. You will only benefit by having a kit readily available to you, should the need arise. If you already have a kit in the house, the job is done for you. If not, we will compile a small list of necessities. First, you will need bandages. This doesn't necessarily mean that you need the branded, sticky type. Any sort of clean, cotton cloth will do. Use clean t-shirts or sheets and rip them into strips that you can use to wrap around a wound. After you have created ten or so bandages, wrap them in another clean cloth. This will make sure that your bandages stay clean, which is of the utmost importance. Next, you will need to find something to use as an antiseptic. Rubbing alcohol, hydrogen peroxide, and liquor will also serve as suitable antiseptics. Try to avoid any liquor which has sugar already mixed into it, as this can cause the wound to possibly fester. For the absolute worst cuts and wounds, it might be helpful to have a way to suture the wound. If you can find a sewing needle and fishing line, this will work almost as good as a doctor to close the wound, provided you do it correctly. Go ahead and put the aspirin and other OTC medications you found with this kit, that way, all of your health items are in one easy to reach place.

Now that your utilities are shut off and your house is free from the dangers they possess, we can look to actively fortifying your home for both safety and sustainability.

FORTIFYING YOUR SHELTER

While fortifying your home, there are a few things to keep in mind. First, one must remember that without modern appliances, temperature regulation becomes an extremely important issue to manage. At night, temperatures dip below freezing in many parts of the world, doubly so if it is winter. This puts you at a great risk for hypothermia, and steps must be taken to prevent this. On the opposite end of the spectrum, extreme heat can cause heat stroke, which could lead to death. Secondly, you will recall that not only are you defending against natural threats, but also against possible violence perpetrated by fellow citizens. You will have to make sure that your structure is not only fortified against damage from nature, but also against unwarranted entry and safe from looters.

Barricading The Windows And Doors

The weakest points of your house, both in regards to nature and looters, are your glass windows. Windows made of glass are at an extreme risk for breaking, either via debris thrown about in the wind or by people trying to enter your home. Once this glass has broken, the glass can never be repaired, only mended. This means that you can lose precious heat and insulation integrity through essentially a hole in your wall, or, people will be able to enter unchecked. When it comes to fortifying your home, one of the first things you should do is to board up the windows and ensure there is proper insulation between the layers. Before doing so, identify two windows that you wish to be your "lookout" windows. These windows will provide you a clear vantage point to seeing both the front and back of the house, so you are aware at all times of your situation. Go to the rest of your windows, and in between the blinds and the actual glass, fill it with as much paper and other insulation as you can. Use ripped up books, junk mail, or newspaper to do so. Cloth scraps and rags also work well in this regard, basically anything that can provide a barrier that can trap either the heat or cold. This will provide a layer of insulation that can help prevent heat loss from within your home should temperatures drop below freezing. If you live in a warm climate, you will have to make sure that you leave enough windows for proper ventilation and cooling. However, boarding up your windows is still important, for the reasons we spoke of before.

You will need to barricade these windows with something strong and sturdy. There are multiple ways to do this, but the easiest way is to salvage and use the furniture within your house. Wooden tables, cabinet doors, shelves, and entertainment centers can all be harvested and broken down for their wooden boards. Use these boards to barricade all the windows you can reach. Remember those two windows that you marked as vantage points earlier? Barricade these windows as well, but leave enough room on the side so that you can see out clearly, but still have a degree of protection. Being able to see the outdoors will allow you to be aware of your situation at all times, and spot threats as they arrive.

Doors within the house must also be barricaded, however they should be done so in a way that allows access out for those within Use the heaviest furniture you have that is easily slideable, such as a couches, mattresses, or large dressers, and push them up against all the doors in your house. If someone does try to enter, they will have a difficult time due to the blockage behind the door, giving you precious seconds to act. Additionally, you can enter and exit at your own whims, moving the furniture out of the way when necessary. It is important to make sure that you can do this quickly, as should you need to leave your house in a rush, you cannot risk precious time unable to reach the door. Also keep in mind any other members of the household. Should you become incapacitated, it is important to make sure that everyone has a way that they could exit the household in a rush.

Look For Damage

Now that the windows are barricaded, and you are somewhat safe against any immediate fallout from the disaster, it is important to make an accurate survey of your shelter. If you have chosen to stay put in your house to ride out the disaster, at least for the time being, you must see it as a place of shelter and sustenance. That means that anything you have within the house, can and must be used for survival and advancing your situation. The tools a house can potentially provide are invaluable, and the shelter it provides even more so. Therefore, it is important to make sure that the integrity of your structure is still intact, as your survival is dependent on it. After a disaster, it is possible that your house may have sustained damage during the event. This damage can sometimes be very overt and visible, but often times it is not. Less obvious signs of damage can include fine cracks along the walls or structure of the house, which can then spread and become more significant. Make sure that any obvious holes within your shelter are patched, along with any broken windows, even if it is done crudely. If you have an accessible basement, go down and try to assess if there has been any structural damage to the foundations of your house. It will be of no use if it is on the brink of collapse, and if it is, you will have to vacate the house. This is a decision you will have to make while you go around fortifying your home. At a certain point, if you are making too many repairs, it is possible that your home is not sustainable. However, as long as it has not suffered significant breakage or flooding, the house should be safe to remain in.

Staying Warm

You will need to then decide where you and your family will be sleeping. While the bedrooms are an obvious choice, sleeping together in one room can prove more efficient in terms of safety and bodyheat. This can vary depending on the size of your household, but generally a central location is usually the safest. Another thing to keep in mind if you live in a cold climate is heat. It will be necessary at night to build a fireplace, so for homes that have one, the most ideal place for sleeping would be beside the fireplace. If you don't have a fireplace, but will be facing freezing weather, you can make an improvised fire pit. Something as simple as a large metal pot will suffice for a very small, controlled fire and stove. Even the smallest fire can provide life-saving heat, especially if radiated correctly.

Go through your house and collect all of the blankets and sheets you can. These can provide invaluable insulation, and in more temperate climates, even eliminate the need for a fire. It is important to remember that you will lose most of your body heat through the ground, so padding below you is almost more beneficial than above you. Pile all of the blankets and pillows in the location that you will be sleeping, and arrange a sort of pit that will face the fire. Be certain to make sure you are in a safe distance away, but essentially your "bed" of blankets and sheets will serve to trap and conserve both body heat and radiated heat from the fire.

Outdoor Security

Before finally sealing your house, it is smart to add some basic security to the outside. To do this, there are a myriad of different traps and tripwires you could deploy, however, we will stick to a basic tripwire. To make it, all you will need is two items to act as stakes, and enough cord to cover an area horizontally, with enough line left to make it inside to act as a signal. For cordage, parachute cord or something similar is ideal. Fishing line could also be used. Simply stake off a horizontal area close to your door, and tie the wire across the two stakes. Run the rest of the wire very tightly into your house, and tie it off to an object heavy enough to keep the line taught, yet light enough to be moved if the line is disturbed. This will act as a rudimentary signal should someone try to enter your property. Ideally, you would want to set traps with a trigger system that would set off a mechanical or electronic alarm, but that is a more advanced technique. In a pinch, the method outlined here can potentially buy you at least a few sections of preparation, which can prove invaluable.

Now, your home is safe from both the inside and outside, and from natural and physical threats. At this point, a couple of hours will have undoubtedly passed since the time of the disaster, and we will have to look to the concerns of sustaining the body in terms of food and water.

EXTRACTING WATER

Water is without a doubt the most essential element when it comes to survival. Without water, dehydration sets in and you die a slow death within three days. For comparison, you can survive for more than three weeks without eating any food, surviving only on the nutrients already stored in your body. The risk for dehydration is elevated in both extremely hot and cold environments, which introduces an added stress on the body and drain more moisture. Side effects of dehydration sickness can include headache, dizziness, nausea, and confusion. If left untreated, you slowly lose your mind as your body shuts down all of its organs. The average adult needs about 1.8 liters, or half a gallon, of water a day for comfortable survival. This figure can be stretched during a survival situation, but eventually a full rehydration will be necessary. As such, securing and rationing your water supply is one of the first things you should do upon making sure your home and territory are secured. Fortunately, there are places in your home where you can gain an additional, and quite substantial water supply.

Before we discuss extracting that water, we will talk about a common method that is taught in many survival guides, but that I disagree with. Many guides advocate filling your bathtub with water as soon as a disaster strikes, ensuring that you have a plentiful supply of water. While this certainly provides a large amount of water, it is not guaranteed that the water is safe to drink, even if treated. After a disaster, it is possible that the water lines may become compromised. If you fill the tub right after the disaster, it is possible that some of the water coming through is not safe for consumption. There is no real way to tell if the water is free of chemicals. Due to this, even using the water for cleaning can provide a certain degree of risk. Should chemicals or parasites in the water enter your blood stream through your eyes or mouth, you will fall victim to it just as if you had ingested it. Variables like this are not your friend in survival situation, so it is best to keep the variable entirely out of the picture.

The Toilet Tank

The first, and by far the easiest, place to get fresh water in your home is your toilet's water tank. While some people may be offended by this, and consider it unsanitary, the fact is that the tank is never subject to any of the areas where waste is dispensed. This water, unless you have added chemicals yourself, is fresh and safe to drink. The modern, low flush, toilet tank holds about 2-3 gallons of water in its tank. Older toilets can hold up to 5-7 gallons of water in their tanks. That is enough water to sustain one person for nearly four days, and easily accessed with little to no calories expended. Although this water is certainly safe to drink as is, I recommend treating the water, just to err on the side of caution. We will discuss methods of purifying the water at the end of this section.

The Hot Water Heater

The most substantial location to extract water from your house is your hot water heater. Provided that you shut off your water promptly after the disaster, as the guide outlined, this water is safe and isolated from the main water line. An average hot water heater can hold anywhere from 20 to 50 gallons. As you can see, this is an invaluable asset for safe drinking water within your home. This water can potentially keep you and your family hydrated for weeks, if not at least a month. While each model of hot water heater differs slightly when it comes to extraction, for the most part the directions are very similar, and will be outlined here. First, make sure that your heater is unplugged from the wall outlet. Even if the power is turned off, as it should be, it is possible that you can damage your water heater when the power is turned back on. Essentially, the element will burn itself off, leading to costly repairs in the future. Go ahead and unplug this now, just to ensure that it is not forgotten about later. Depending on how long the power has been out, allow extra time for the water inside the heater to cool, as it can be extremely dangerous to work with the water while it is still heated. Not only can the water be scalding hot, but also the tank can be pressurized due to the steam of the water. Once enough time has been given to allow it to cool, you will need to relieve this pressure from the hot water line. The easiest, and most universal, way to do this is to turn on the hot water from the tap closest to the heater. This will effectively bleed the line and reduce the pressure within it. Alternatively, you can remove the actual hot water line directly from the tank. Finally, you are ready to extract the water from the drain valve at the bottom of the tank. The initial outpour of water may be murky; this is due to sediments that collect at the bottom of the tank. This is normal, and you should simply discard the first few gallons of water to get rid of this. If there is still a bit of sediment in the water after the initial gallons, let the water sit in a container and the sediment will settle on the bottom of the water, allowing you to siphon out the clear water from the top. Extract as much water as containers you have in the house to store it in. If you have a readily available supply of water conveniently located, you will be better prepared for any further emergencies that may arise.

Treating The Water

Now comes the process of treating the water to ensure that it is safe for consumption. I recommend that you treat any water that did not come out of an unopened bottle, before you consume it. While the water in the hot water tank is certainly safe to drink, it is always better to be safe rather than sorry. If you catch a sickness from the water, you face the risk of becoming even more dehydrated at a more rapid rate via dysentery. The easiest way to disinfect water is by using simple household laundry bleach. Put six drops of bleach per every gallon of water, and let it sit for thirty minutes. This will effectively kill any parasites or bacteria that were in the water. If bleach is not available, you can kill the bacteria by utilizing heat. Put the water over a fire, and allow it to come to a rolling boil. This means that the water is visibly bubbling and boiling, but not overly so. This allows you to maximize the water you have, and not lose so much to evaporation. Let the water do this for about ten minutes, and after that, it will be safe to drink, effectively pasteurized.

Rationing

Finally, you will need to ration the water out in order to ensure that it is being used as efficiently as possible. To do this, you will need a way to measure out the water. The most efficient way to do this is to use a container that is premeasured, such as a milk or juice carton. Put whatever liquid is in the container in cups to save for later, and using a small amount of water, clean and rinse out the container as best as you can. As you will be dipping this into your larger supplies of water, you don't want to compromise the larger water supply by introducing sugars or dairy to it. Now, the amount you can ration out depends solely on how many containers you have at your disposal. Ideally, you will want to ration out your water by days, even if it is only a couple of days at a time. This way, you can prevent waste or overdrinking, while your larger water supply remains safely in wait. Remember, the formula we will be using is ½ gallon of water a day, per person. If you have soda bottles, this equals out to roughly a little bit less than a full 2 liter bottle. So, take however many people you are trying to sustain, and ration out a half of gallon for each person.

Again, make sure that the container you are using is easily covered, in order to keep the water clean. If you lack larger storage type containers, then just use smaller containers and refill them as necessary. Remember that you still have juice and milk from the cartons that you emptied for measuring. Drink this liquid first, as it is perishable and will go bad without refrigeration. No matter how you ration it out, it will be better than blindly drinking from the main supply. In extreme situations, you will be surprised at how fast the water depletes, and if that should happen, you will have to leave your shelter and look for other methods to procure water.

RATIONING YOUR FOOD

As I mentioned earlier, there have been cases of people surviving more than three weeks without any food. This, however, is an extremely dangerous feat, and doubly so when coupled with a high stress situation in a possibly extreme environment. While the body will persevere, adverse effects from starvation can develop in as little as two days. Extreme stomach cramps, nausea, and headache are just some of the discomforts you will feel without food. This is the last thing you need in a survival situation, where keeping your mental faculties clear is of tantamount importance. Chances are that within your house, you have a substantial amount of food stored away. This food, however, is probably spread out all throughout your kitchen in different cabinets and cupboards. As we did with the water, it is important to know exactly how much food you have at all times, and how much food you have to begin with. Everything must be measured, surveyed, and accounted for in a survival situation. In this section, we will focus on how to make sure you are using your food supply smartly and responsibly.

The Fridge

First, we will focus on the food that is most at risk for spoiling – the food in your fridge. You will want to use as much of this food as possible during the first stages of the disaster. For the most part, the food in your fridge will be the most nutritious and calorie rich sustenance available to you at the outset of the disaster. You will need to separate the items in your fridge into two categories, vegetables and protein. Any vegetables in your fridge will be able to last for at least a few days outside of the fridge. As such, this food can be stored to be used sparingly with other meals, to ensure you are getting proper nutrition while surviving. So, take all of your vegetables out of the fridge and store them somewhere for the time being. Most fridges have vegetable drawers that can be removed from the fridge, and this makes for a great and easy storage container. Now, you will be left with only the most perishable goods in your fridge. This is where smart rationing is integral to maintaining efficient use of your food supply. The most perishable items will need to be consumed first, and done so before they spoil. Without a working fridge, meat and dairy products will quickly go bad. Therefore, most of the meat you have in your fridge will need to be cooked right away, and any yogurt or dairy products consumed. The FDA recommends that both raw and cooked meat only be left out for a maximum of two hours. In a survival situation, this is not a feasible amount of time for it to be left out. You will have to leave the food out for a bit longer, but the faster it is cooked and consumed, the better. Eggs can be left outside of the refrigerator safely for an upwards of two weeks, so these can be an excellent source of fats and proteins which can help you later on down the road.

Cooking The Meat

Now that your stove is not working, cooking the meat you have can prove a bit difficult. If you do not have access to a propane cooker or hot plate, you will need to improvise a stove. If you have a fireplace, that's where you should be cooking. This is simply because the chimney can provide an exhaust for the smoke that your fire will produce, which will be a considerable amount. It is important to make sure you have proper ventilation when cooking, as filling up your home with smoke is dangerous. Try to do this at night, so the smoke produced from the fire does not attract any sort of unwanted attention. Also make sure that the imminent threats of the disaster have passed, and that you will not make yourself vulnerable by cooking. If you don't have a fireplace, then you can create a small stove in a solid metal pot. Try to do this in the garage if you have one available, as the concrete surfaces and access to the outdoors can make ventilating it easier. Make sure that this pot is not coated with any Teflon, and is purely metal. You can start a fire in the pot, and use the flame and coals to cook your meat, purify water, and even roast your vegetables. Cook and ration this meat out equally for everyone, and make sure that it is eaten sometime within 12 hours of its cooking. Having an excess of food at the outset is not a bad thing, as if you run into problems later, you will at least have had a good amount of nourishment. The more fat you can carry on your body during this time, the better, as should you run out of food your body will first metabolize this excess fat.

The Freezer

Items in your freezer can prove a bit more difficult to deal with. For at least the first few hours after power is cut off, your freezer will remain cold. You can utilize this to help keep meat and dairy from your fridge cool for a little while longer, giving you more time to tend to other duties. However, make sure to keep a watchful eye on this, as the freezer will increase in temperature rapidly without power. The food stuff within will be at an extreme risk to rot quickly if it is left in a warm, enclosed space. For the frozen items in your freezer, you will have to make a decision on a case-by-case basis. Any meat that has been frozen can be slowly thawed and cooked when properly defrosted. Any sort of frozen, precooked food, such as chicken nuggets, French fries, or chicken wings might prove more trouble than they are worth. Many of these foods depend on being cooked at high temperatures in a conventional oven or microwave. Thawing this food can also pose a problem. Cooked foods are safe to heat after being thawed, however, they must be slowly thawed in a cool place like a fridge. Now that your fridge is out, you don't really have anywhere to thaw the meat out safely. It will be susceptible to bacteria, and if you get sick from it, you face the ever-imminent risk of dehydration. If your other food sources are lacking, eating these goods will become more important. However, if you can spare to not eat it, then do so.

Canned Goods

Now, go through your cupboards for bread and canned good. These two items will be the most valuable foodstuffs you possess, because for the most part, they last a very long time and do not have a set expiration date. Create a pile of all of the food in jars or cans that you can find. This can include things such as soups, canned vegetables or fruits, tomato sauce, tuna or even peanut butter. All of these items are extremely nutrient rich, and will last indefinitely if sealed. Any bread in your house will also be a huge boon to your survival. Bread can last for a couple of weeks, at the very least, without and refrigeration. It is only unsafe to eat bread if there is visible mold on the bread, so keep a watchful eye for that. Take a count of all of the canned and jarred food you have. This is the food that you will want to be rationing the most carefully.

Rationing

Ration out the food in daily portions, according to how many people you have to feed. To do this, you will need to read the nutrition labels on each of the can, so you can ration it out according to calories and nutrients. Roughly speaking, 1000-1500 calories a day will keep you feeling decently full. This amount can be dropped even lower, but some discomfort may be experienced. My advice is to ration out your food at or below 1000 calories per person, per day. Remember, you have meat that you will be eating for the first couple of days at least, and if you have eggs, those can stay fresh for more than two weeks. With all of the jars spread out in front of you, it becomes easier to see what you have to work with. Using 1000 calories as your rough baseline, ration out how many days worth of canned goods you have. It is important to keep in mind a balanced diet whilst rationing out your food. Remember, you also have vegetables that you put aside earlier that can be eaten in conjunction with the canned foods. Make sure that you spread out the different varieties of canned goods throughout the days. Try to include at least one protein, one fruit, one vegetable, and a bit of bread each day. If you are low on any of these, you can compensate by adding more vegetables or bread.

You should monitor your food supply closely for the first couple of days, and make adjustments to the rationing accordingly. If food is not being used as much as you expected, stretch out the rations to last you longer. If you find that you are still marginally hungry, perhaps try adding a bit of food to the daily rations. Either way, rationing is important for the simple fact that it gives you a set time as to when you will run out of supplies. This can aid you in your planning of what to do next, and how long you can afford to stay within the shelter.

STARTING A FIRE

Many of the techniques we have discussed in the guide thus far have depended heavily on the use of fire. After shelter and sustenance, fire is the most important element to survival, enough so that it warrants its own section. Fire can allow you to warm up your shelter, purify your water, cook your food, and even signal for rescue. Knowing how to create a fire is extremely important, and is not as easy and straightforward as it seems. It is not simply rubbing sticks together, or using a lighter to set a piece of wood aflame. Starting a fire takes three essential components, all working in perfect tandem with one another to get the initial blaze going. That is why I will go over the basics of fire making, and the components you need to collect.

Clearing The Area

Before you start a fire, you will first need to consider the area in which you do so. Fire is an extremely volatile substance, and can spread quickly, reducing your shelter to ashes and possibly leading to death. If you have a fireplace, that is the best location for the fire. It has a built in exhaust, which can vent smoke and fumes, and is made of brick, which cannot combust. If you do not have a fireplace, you can make a small fire using the method we talked about in the cooking section, a large, all metal pot. If it gets extremely cold, you can make multiples of these small stoves. Ventilation will be a consideration you will have to make when you have an active fire. Filling the house with smoke can be dangerous on the lungs and induce suffocation. For cooking, which will produce the most smoke, a location like the garage is the most convenient. A garage can be easily ventilated, and is not at risk for combustion due to being made from concrete. Inside the house, try to position the location for a fire near an easy accessible window that you have not completely boarded, which can allow for proper ventilation.

Elements Of The Fire

To create the fire, you will have to find something that can produce a spark, or even ideally a flame. Lighters are obviously the best fire-starting tools, and if you have one, a fire is nearly guaranteed. However, if you don't, there are methods of creating fire from items around your house, which we will discuss later. The three elements of a fire are the tinder, kindling, and fuel. The tinder takes the initial spark to produce a small flame, which you then use to burn the kindling, and keep it going with the fuel. The tinder will be your most combustible material, and has to be because it will act as a medium to transfer the energy from your spark to the kindling. Some of the absolute best tinder you can find in your house is dryer lint. Dryer lint is extremely fibrous, and as such, there is a lot of surface area in which the spark can catch to produce a flame. Usually, one spark from a flint will be enough to light the lint. Toilet paper can also be used as tinder. The trick is to rip up the toilet paper to give it more surface area so the spark can catch better. The more jagged edges it has, the easier it will be set aflame. For both of these methods, bunch them up until they resemble a bundle, or more accurately, a tinder bundle. For your kindling, you will use wood pieces that are large enough to produce a decent flame, but small enough to catch fire from the tinder. Arrange your kindling in a way that will provide oxygen through all parts of the fire. The best arrangement is a teepee arrangement, where the sticks are vertically pushed to the center out, resembling a teepee. Make sure to leave room in one of the sides, so you can place the tinder inside of the teepee. Additionally, this design will provide a sort of outlet for the smoke near the center of the teepee. Because of this, you can decide where to place the fire to best ventilate the smoke. The fuel will be the larger pieces of wood you have that keep the fire going. Keep in mind that you will constantly have to be tending to the fire, making sure that the flame is going and that there is enough fuel within it. To recap, light your tinder bundle, place it inside the kindling teepee, and let the flame get going. Then, add your fuel as necessary to keep the fire alive.

On Firewood

You will want to make sure you have an ample supply of wood to keep your fire going. Always make sure that you build your fire to a manageable size, as a large one can expend fuel extremely rapidly. As far as wood, natural wood is the best to use for fires. If you have access to any trees or bushes outside of your house, gather an ample of supply of that to bring in, especially before you settle in for the night. However, if your environment is completely urban, you will need to utilize the wood that can be found inside of your home. First, check your garage. Usually, people will have a couple of planks of wood that they did not even know about in the garage. This wood can be broken down according to the sizes you need. One plank of wood can be used as tinder with its sawdust, kindling with smaller splinters of wood, and fuel with the larger pieces. Essentially, one plank of wood and a spark is all you need to start and keep a fire going, at least for a short time. Wood from your furniture can also be broken down to be used as fuel. This option, however, does pose a danger. Often times the wood in furniture is treated with different varnishes and lacquers. These chemicals can produce dangerous fumes and smoke when burned. It is best to stay away from wood that looks visibly treated, and if you do use it, make sure that you have adequate ventilation and only use it in small amounts for warmth.

Improvised Firestarter

If you do not have a lighter or other flame producing device, here are two fairly easy methods to produce a flame using items that you can find around your house. First, is the battery method. A double A battery works best for this method. Essentially, you will short out a wire using the two terminals of the battery in order to produce and utilize the heat within that same wire. Perhaps the easiest way to do this is to use a gum wrapper. Rip the gum wrapper so that the center of it is the thinnest part. You need this part thin so it can heat up rapidly, and produce a flame once touched to the terminal. Get your tinder bundle ready, attach both ends of the gum wrapper to the positive and negative terminals of the battery, and touch the thin, center part to the tinder. The electricity in the battery will be enough to heat the wrapper to combustion. If you don't have a gum wrapper, you can use any thin conductive piece of metal or wire. The amount of time it takes to heat up will depend on its gauge, so it is best to find something very thin. A second method is one that most people are familiar with, utilizing a lens. This method only works if you have visible sunlight available in your environment, which might not always be the case. A magnifying glass works best for this, however, any sort of convex lens can be used to produce the same effect, including a camera lens or a pair of glasses. Provided you have sunlight, you can utilize this method as much as you need to, over and over again. Position the lens so that it focuses the sunlight on as small an area as you can manage. Keep this beam of light focused on your tinder, and eventually, it will heat up enough to combust.

Once you have started an initial fire, you should never have a reason to start another one utilizing the methods above, provided you stay at the same shelter. You can use the coal and burning embers from the main fire to light any other fires you may be starting. Simple take a burning stick or pile of embers, and apply them to your tinder bundle. Gently blow on the ember, allowing it to spread due to more oxygen. Eventually, that tinder bundle will light aflame and you will be ready to start another fire.

GATHERING YOUR TOOLS AND WEAPONS

At this point in the guide, you are in a fairly good position, at least for the short term. Your house has been fortified and secured both from natural threats and from looters. You have gone through your house and made sure there are no weak points in its structure. Additionally, you have extracted all of the water out of your house for drinking, and rationed both your food and water. The only exigent factors that could compromise your safety are further disasters, or extremely violent looting. As such, you must prepare for both of these threats by making sure you have the tools necessary for sustenance, as well as weapons to defend yourself should the need arise.

Finding Tools

If you have a box of tools somewhere in your house, go and grab it. Take a survey of what is inside. If you have access to nails, these can prove extremely helpful to bolster your existing fortifications. You will be able to nail in place the barricades that you put up earlier, and create a more secure and permanent fortification. The next thing you should secure would be a good knife. Knives have many functions in a survival situation, they can be used for cooking, eating, cutting wood, cutting rope, cauterizing wounds, and hammering, amongst other functions. A good blade is one that is thick enough to perform all of the activities above, but also very sharp. A dull knife is extremely dangerous, as you exert more force to cut and can damage yourself in the process. Go through the knives in your house and pick the ones that are best suited for these activities. Make sure that you have a few backups as well, to do the rougher work. Ideally, you will want to only use your main knife when you need a blade that is sharp, so you can preserve the edge on the tool.

While you are searching for tools, take a moment to check on the status of your automobile, if you have one. It can prove an invaluable asset should you need to escape quickly at any moment. Make sure that the car starts properly, and runs well with no leaks. In a pinch, if you start to succumb to the cold or even the heat, you can utilize your car's temperature control system to regulate your own body temperature. If this is done, make sure that your garage is ventilated and that you keep an eye on your fuel gauge. By the end of this situation, your car can potentially be the most valuable tool that you have.

On Firearms

Finally, it is time to turn our attention back to the human threat. Should supplies run scarce, it is likely that people will began to scavenge and loot neighborhoods for supplies. Violence is the absolute last method of recourse you should take in a survival situation, but should it be thrust upon you unwillingly, you must start thinking about how you will protect yourself. If you do have any firearms within the house, make sure they are completely secured, and in a location which only you can access easily. Additionally, make sure that you have one on you at all times, preferably one that can be easily concealed, and that you have extra magazines for. Take a count of how much ammunition you have available to you, and be aware of that count at all times. Any extra magazines you have should be reloaded with ammo, so they can be used quickly should the need arise. Follow basic firearm safety, such as keeping the guns away from children, making sure the safety is always locked, and never pointing the gun at anything you do not intend to harm. If you do indeed have a firearm, you are at a considerable advantage when it comes to threats. Any sort of blunt or sharp weapon pales in comparison to a firearm. Remember, if you pull the firearm, you should be ready and have intent to use it. Often times, the person who leaves a gunfight is the one who reacted the most quickly. However, this does not mean that you are not at risk simply because you have a gun. Depending on the amount of people you will be facing and their experience level, it is still possible to lose an altercation when you have a firearm, especially if you are caught by surprise. Firearms are not always fool proof, and are known to sometimes have malfunctions such as gun jams. It is smart to have an arsenal of melee weapons, should your firearm fail or more importantly, if you do not have a firearm to begin with.

Improvised Weapons

The best type of weapon will be one that keeps as much distance as possible between you and your assailant. These types of weapons can be used both defensively and offensively, and work relatively well to keep you out of range during the altercation. Additionally, these weapons can be modified to make them more lethal, with relative ease. Let's take a look at items around your house that can be utilized for weaponry. The most obvious choices are larger gardening tools, such as a shovel or a pickaxe. These items have thick, wooden shafts with metal at the end, which can be used to swing or thrust with lethality. The downside to these weapons is that they tend to be quite heavy, and thus unwieldy in close quarters. Chances are that if it comes to a physical altercation, it will be defending one of the entrances to the shelter and be taking place in quite close quarters. As such, it would be more prudent to collect weapons that are more easily handled in an indoor setting.

A broom can be improvised into a weapon by attaching a knife to the end. A metal pipe or even baseball bat can provide enough concussive force to down an opponent. Basically, anything that is long and fairy dense can be used as a weapon in a pinch, or be improvised to become a weapon. It will be helpful to have an assortment of options to choose from, if the need should arrive. Remember, I cannot stress enough that avoiding violence is always the best course of action. A fight introduces too many variables to the situation, which you cannot control, and you are almost certain to sustain at least one wound following the altercation. However, if the choice is between abandoning your shelter and supplies or fighting, it might be worth it to stand your ground. Make sure that you have weapons near every entrance or main window to the house, so they can be easily reached. However, do not put them in a place where a possible assailant could also reach them. You want to make sure that everyone in the house knows the locations of these weapons, and are able to reach them without alerting the assailants to their location.

CONCLUSION

All of the steps in this guide can be done in just a few hours. If you have multiple people in your home, it is important to make sure that everyone is pulling their weight. Delegate out simple tasks such as the gathering and collecting of materials, while you take care of the more complicated ones. Use every asset you have to ensure you and your loved ones make it out of this alive. Additionally, you need to remember that your home has now become your shelter. Many people grow attachments to the things inside their homes, and will be unwilling to salvage them for survival. The most common reasoning for this is that they do not believe in the severity of the situation, or put too much faith into being rescued. Often times, this line of thinking is extremely dangerous and can lead to death. You must realize that while items may have sentimental value, everything inside your home is now there strictly to keep you alive. Burn books for heat, if you have to. Tear down furniture, put nails in the wall...all of this will be insignificant in the long run, and making it out alive is more important.

Now that you have read this guide, you have a very good idea of what to do following the events of the disaster. All of the tasks outlined in this book can be completed in just the course of a day. This means that within a day from the time of the disaster, your house should be secured, as well as your water and food supply figured out. Not only that, but you should have a good idea at how long you have before you need to start seeking out other options in terms of sustenance or even shelter. If your shelter is being pounded by nature, and losing its integrity, then you will eventually have to make the decision to either affect rescue or occupy another shelter. Additionally, if your food, or more importantly, water, supply starts to run low, you will have to procure a different means of attaining it. Often, this will mean leaving the safety of your shelter to scavenge and search for yourself. In further guides, we will discuss methods in which this can be done safely, and methods to turn your shelter back into a permanent place of safety.

However, the most important takeaway from this guide is you now have the ability to remain calm throughout the situation. Instead of panic gripping you, because you are unprepared or unsure of what to do, you can now look at the situation logically. In decisions of life and death, always make sure that logic is your primary reasoning. Look at the situation by breaking it down into mental steps. Shelter, Water, Food, Fire, Weapons. That is the order you should tackle your tasks in, in the order of most importance. Stay calm, act fast, think logically, and you are sure to survive any disaster.

Finally, if you enjoyed this book, please take the time to share your thoughts and post a review on Amazon. It'd be greatly appreciated!

Thank you and good luck!

www.ingramcontent.com/pod-product-compliance
Lightning Source LLC
Chambersburg PA
CBHW070802180526
45168CB00004B/1729